THOMAS CRANE PUBLIC LIBRARY
QUINCY MASS
CITY APPROPRIATION

The Ice Ages

by Don Nardo

KIDHAVEN PRESS
An imprint of Thomson Gale, a part of The Thomson Corporation

Detroit • New York • San Francisco • San Diego • New Haven, Conn. • Waterville, Maine • London • Munich

© 2006 Thomson Gale, a part of The Thomson Corporation.

Thomson and Star Logo are trademarks and Gale and KidHaven Press are registered trademarks used herein under license.

For more information, contact
KidHaven Press
27500 Drake Rd.
Farmington Hills, MI 48331-3535
Or you can visit our Internet site at http://www.gale.com

ALL RIGHTS RESERVED.
No part of this work covered by the copyright hereon may be reproduced or used in any form or by any means—graphic, electronic, or mechanical, including photocopying, recording, taping, Web distribution, or information storage retrieval systems—without the written permission of the publisher.

Every effort has been made to trace the owners of copyrighted material.

LIBRARY OF CONGRESS CATALOGING-IN-PUBLICATION DATA

Nardo, Don, 1947–
 The Ice Ages / by Don Nardo.
  p. cm. — (The KidHaven science library)
 Includes bibliographical references and index.
 ISBN 0-7377-3055-2 (lib. bdg. : alk. paper)
 1. Glacial epoch—Juvenile literature. 2. Climatic changes—Juvenile literature.
 I. Title. II. Series.
 QE697.N37 2005
 551.7'92—dc22
                                                            2005001695

Printed in the United States of America

Contents

Chapter 1
Memories of a Frozen World 4

Chapter 2
Earthly Causes for Ice Ages 13

Chapter 3
Cosmic Causes for Ice Ages 23

Chapter 4
Global Warming and Future Ice Ages . . . 31

Glossary 41

For Further Exploration 43

Index 45

Picture Credits 47

About the Author 48

Chapter 1

Memories of a Frozen World

Long, long ago, large portions of the world that are now warm and green were buried under vast layers of ice and snow. On several different occasions, huge **glaciers**, slow-moving rivers of ice, crept across the land. These heavy, powerful glacial giants leveled forests and sliced off the tops of hills. Not surprisingly, temperatures in the affected regions remained below zero all day, every day. And all the while, living things either retreated before the oncoming ice or learned to live in the cold.

Modern scientists appropriately call these cold, bleak periods of the past "ice ages." But scientists were not the first people to suggest that the world had once been much colder than it is now. Scattered tribes of humans lived in the last ice age. And as the great ice sheets finally retreated, memories of the frozen world remained. Though they grew dimmer over time, the memories continued to pass from generation to generation in the form of folk tales. A few of these stories survived into early modern times in remote parts of Europe and elsewhere.

When he was a young man, Jean de Charpentier (1786–1855), a German-Swiss geologist, heard these folk tales. Fascinated, he came to believe that they might be based on fact. So he closely examined many of Europe's mountains and valleys. Between 1825 and 1833, Charpentier gathered a large amount of firm evidence showing that enormous ice sheets had once covered Europe. Other scientists soon verified that he was correct. And they found evidence for similar glacial advances in many other parts of the world.

An enormous glacier creeps across a frozen wilderness. During the ice ages, glaciers covered much of the Earth's surface.

Early Ice Ages

In fact, in time scientists discovered that there have been four major ice ages, three of which occurred long before humans existed. The first, the Cryogenian ice age, took place between 800 and 600 million years ago. Seemingly endless winter covered large portions of what are now China, North America, northern Europe, and Australia. Even Africa, most of which is very warm today, did not escape the cold's grip. What is now the Sahara Desert became a frozen wasteland.

Some of these frigid areas were covered by deeply packed snow. Others were buried beneath massive glaciers. These long-lasting rivers of ice formed from the slow buildup of frozen water. Little by little, a young glacier got thicker and thicker and heavier and heavier. Eventually, its upper layers became so heavy that they crushed and deformed its lower layers, and the force of gravity caused the lower layers to move sideways and outward.

In this way, the glacier began to move and expand. This movement was usually very slow—only a few inches at most

A computer image shows the Earth during the Cryogenian ice age.

The Ice Ages

A retreating glacier usually leaves behind lakes, rivers, and an eroded landscape.

each day. But the movement was relentless. Days became months and the months turned into years. In turn, the years stretched into centuries, and still the great ice sheets advanced across the land. Nothing could stop them. Their enormous bulk snapped even the largest trees like matchsticks. The ice reshaped mountains, reducing many to rubble, and carved out new valleys, which quickly filled with ice and snow.

The world did not stay frozen, however. Just as they had advanced, in time the glaciers retreated. Much of the world grew warm again. New forests

Memories of a Frozen World

grew, some of them in the new valleys the ice flows had carved out. A few thousand years later, however, the cold advanced again. Then it retreated once more.

This cycle of advance and retreat happened several times in the Cryogenian ice age, as well as in the next two major ice ages. One ice age took place between 460 and 430 million years ago. The other brought a series of glacial advances and retreats between 350 and 250 million years ago. Scientists call a warm period between a retreat and the next advance an **interglacial** (meaning "between the glaciers").

The Most Recent Cold Snap

The last of the four major ice age cycles—the Pleistocene—began between 3 and 2 million years ago. So far, there have been about twenty glacial advances and retreats in the cycle. In a sense, each of these periods is an ice age in its own right. Scientists have studied the advances and named each. The Nebraskan glacial period, for example, lasted from about 470 thousand to 330 thousand years ago—a total of roughly 140 thousand years. (Nebraskan is its North American name, whereas in Europe it is known as the Elbe glacial period. Each ice age has several names, depending on the region affected.) Other glacial periods in the Pleistocene ice age included the Kansan (Elster in Europe), lasting from 300 to 230 thousand years ago; and the Illinoian (Saale in Europe), lasting from 180 to 130 thousand years ago.

The most recent glacial advance was the Wisconsinan (Weishsel in Europe). It lasted from about seventy to ten thousand years ago. The greatest extent of its glaciers and snowpacks occurred roughly twenty thousand years ago. At that time, some 97 percent of Canada was buried in snow and glaciers. And an enormous glacier that scientists call the Laurentide covered the northeastern United States in a layer of ice up to 2 miles (3.2 kilometers) deep.

In this illustration, ice sheets cover much of North America during the Wisconsinan glacial advance.

These mighty ice flows created many of the more familiar land and water features in the northern United States. They carved out the Great Lakes, for instance. They sculpted New York's Manhattan Island and Long Island. And during their retreat, they left behind the massive Ohio River system and Niagara Falls.

Life in the Ice

Although the Wisconsinan and other glacial episodes turned much of North America and Europe into wintry landscapes, these regions were far from lifeless. Throughout the Pleistocene ice age, many species of mammals thrived along the fringes of the ice

During the last ice age, human hunters (above) preyed on huge mammoths (left) and other animals.

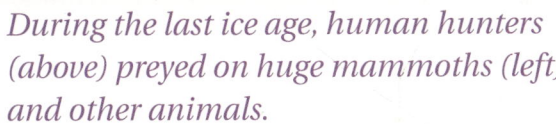

sheets. Some, including beavers, deer, rats, and armadillos, were small. Others were very large.

The largest of all these beasts were plant eaters. Megaceros (meaning "gigantic horn"), for example, was the biggest deer that ever lived, standing more than 10 feet (3 meters) tall and sporting antlers 11 feet (3.3 meters) across. It roamed in large herds across Europe and western Asia

Memories of a Frozen World

until about eleven thousand years ago. A much larger plant eater—the mammoth—also moved in herds. Ancestors of today's Indian elephants, mammoths had long hair and tusks up to 13 feet (4 meters) in length.

Among the meat eaters that preyed on Megaceros, as well as on many other animals, was a giant cave lion. A full-grown adult lion was nearly 12 feet (3.6 meters) long and weighed almost half a ton. The last of these frightening beasts died out only two thousand years ago. Drawings of them have been found in a number of ancient caves.

Of course, these drawings were done by people who lived in the caves. Small tribes of humans managed to survive during the Pleistocene glacial advances. Evidence shows that the tribes were **nomadic**, meaning that they roamed from place to place. Most likely they followed the herds of deer and other animals they hunted for meat and furs. (The people used the furs to make clothes to keep them warm in the cold climate.) The ice age humans also foraged for roots, berries, and other edible plant foods.

Only after the last glaciers retreated—about ten to twelve thousand years ago—did people start to settle down, plant crops, and build permanent villages. Thus, it was the end of the Wisconsinan glacial advance that made human civilization as we know it possible. It is only natural to ask: Will the ice come again?

Chapter 2

Earthly Causes for Ice Ages

Scientists had no sooner discovered the occurrence of numerous ice ages than they began trying to explain how these periods of intense cold come about. They quickly established some basic facts about glaciers, which are major features of all ice ages. First, these ice sheets can come in all sizes. Some are only a few miles long and never grow any larger, while others expand into giants hundreds of miles long and crush nearly everything in their paths.

In particular, researchers wanted to know why and how some glaciers grow larger. In time, scientists noticed something both unusual and disturbing about the growth of these great ice flows. Once they form, they can grow larger and longer-lived of their own accord, without the aid of outside forces. This happens because of a fairly simple natural process involving sunlight, the ground, the air, and the ice itself. Not surprisingly, sunlight plays a vital role in determining if a given climate is warm or cold. When

Scientists study ice layers in a newly formed crack near Juneau, Alaska.

14 **The Ice Ages**

sunlight falls on rocks and dirt, these materials absorb some of the warmth. Over time, they radiate, or give off, much of that heat, warming the air.

Something very different happens when sunlight falls on sheets of ice and snow, however. A large portion of the warmth is reflected away, so the air above the snow and ice stays cold. That makes it more likely that any **precipitation** in the area will fall as snow. The extra snow enlarges the glacier or snowpack. This widens the cold region, causing still more snowfall. As time goes by, the process accelerates. And there occurs a sort of runaway cooling effect that leads to an ice age.

Movement of the Continents

The bigger question is: What causes the cooling and formation of these big glacial sheets in the first place? Numerous theories have been proposed to explain what sets ice ages in motion. Among them are several that involve various physical qualities of Earth's landforms, oceans, and atmosphere.

For example, most scientists think that the movement of the planet's continents has something to do with the ice ages. Only relatively recently did scientists link such movements to the onset of long periods of cold. This is because they did not even know that the continents moved until well into the twentieth century. In 1915 German scientist Alfred Wegener published his now famous theory of continental drift.

If a land mass drifts near one of Earth's poles, large glaciers can begin to form on it.

Earth's great land masses glide very slowly across the planet's lower layers, he claimed, sometimes drifting apart, other times crashing into one another. Because they could not see or measure these movements, other scientists ridiculed Wegener. It took almost fifty years for the scientific community to prove that he was right and accept the theory. A new science was born—**plate tectonics** (in reference to the huge plates of rock on which the continents rest).

In the years that followed, some researchers realized that the processes involved in plate tectonics might at least partially explain the occurrence of ice ages. They pointed out that large ice sheets do not form over open ocean. Any snow that falls there immediately turns into cold water. Land masses are required for the buildup of enough snow and frozen water to create glaciers.

The key factor in the process is the position a land mass occupies on the planet. If a continent lies on

Continental Drift

Scientists believe that continental drift may be partially responsible for ice ages. Here we see how the Earth's continents have drifted over millions of years.

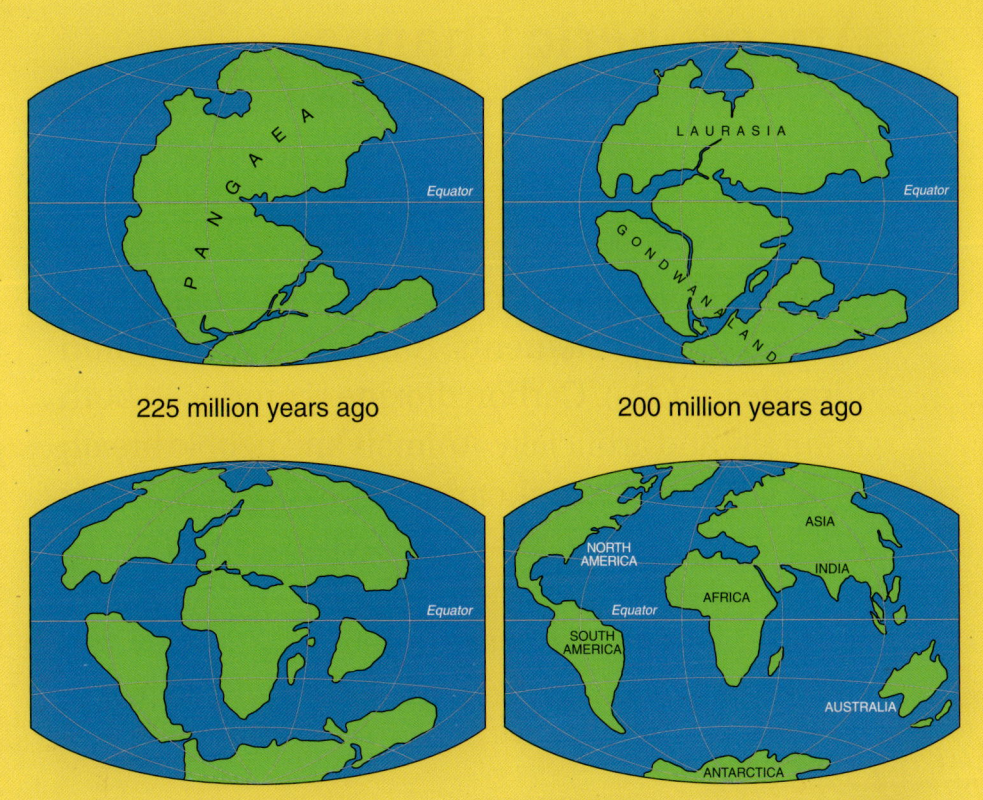

225 million years ago

200 million years ago

65 million years ago

Today

or near the equator, where sunlight is direct and intense, no ice sheets will form on it. But what if a continent lies at or near one of Earth's poles? Sunlight is usually less direct at the poles, so these places tend to have colder climates. If a land mass happens to drift into a polar region, it can begin to build up snowpacks and glaciers. This, in turn, might trigger an ice age. In fact, during the recent Pleistocene ice age, large tracts of land existed within the polar regions. Even today, the ice-covered continent of Antarctica still lies directly over the South Pole. And sections of North America, Europe, and Asia lie above the Arctic Circle (the cold zone surrounding the North Pole).

Atmospheric Changes

Another major theory explaining the ice ages involves changes in the composition of Earth's atmosphere. Most of the atmosphere is made up of two common elements in gas form—nitrogen and oxygen. However, it also contains smaller amounts of several other elements. Among these is **carbon dioxide** (or CO_2). Carbon dioxide is produced both naturally and artificially. Animals and people breath in oxygen and expel carbon dioxide as a waste product. The same gas is also released by various industrial processes and products.

Carbon dioxide is a very potent substance. Even in fairly small amounts, it can absorb and trap heat

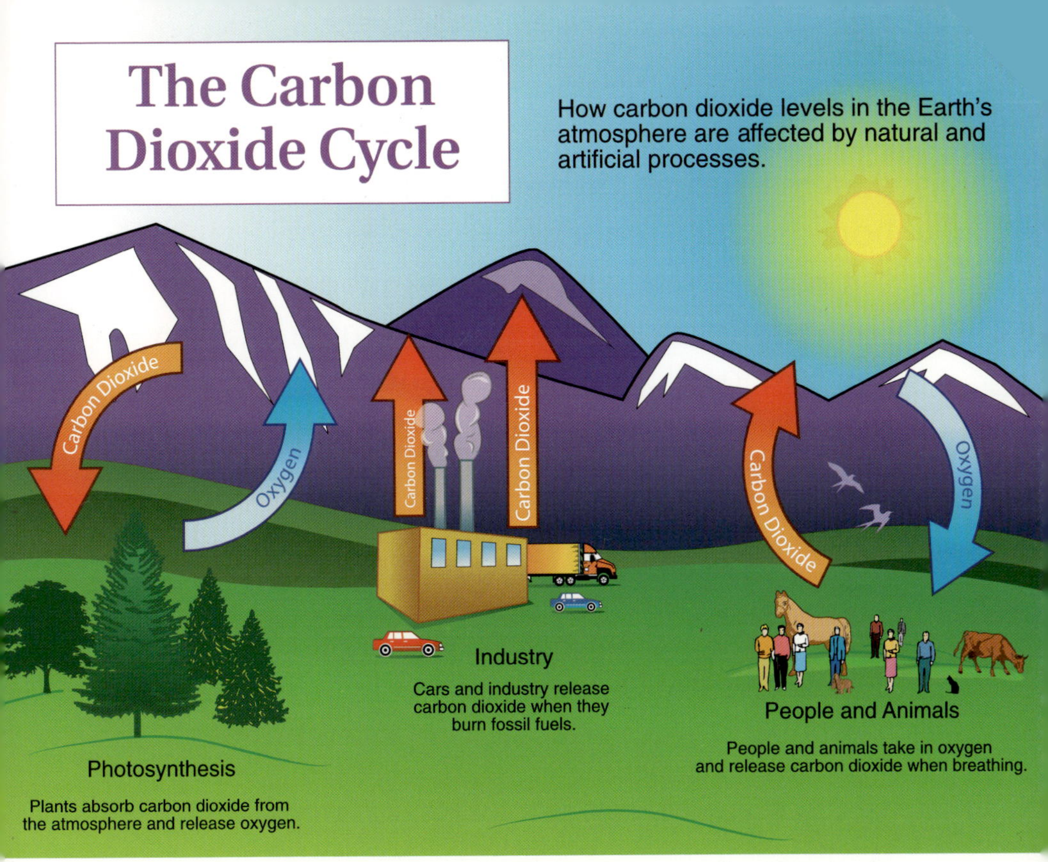

The Carbon Dioxide Cycle

How carbon dioxide levels in the Earth's atmosphere are affected by natural and artificial processes.

Photosynthesis
Plants absorb carbon dioxide from the atmosphere and release oxygen.

Industry
Cars and industry release carbon dioxide when they burn fossil fuels.

People and Animals
People and animals take in oxygen and release carbon dioxide when breathing.

from sunlight. Indeed, the average amount of carbon dioxide in the air helps to keep the atmosphere warm enough to support life on the planet. On the other hand, if the amount of carbon dioxide decreases enough, the atmosphere absorbs less heat. And the air gets colder, especially in northern and polar regions. This, in turn, helps produce snow and form snowpacks and glaciers. Under the right conditions, the result might be an ice age.

A number of processes can contribute to a reduction in the carbon dioxide content of the air. For example, green plants absorb carbon dioxide (and give off oxygen as a byproduct). Some scientists have proposed that on occasion, microscopic

Earthly Causes for Ice Ages

plants in the oceans suddenly become more abundant than usual. They absorb large amounts of carbon dioxide, reducing levels of the gas in the atmosphere long enough to make some regions colder. Other scientists question this idea. They say it is hard to tell whether atmospheric changes make the air cool off or already cool air causes the atmospheric changes.

The Ocean Conveyer Belt

A number of scientists believe that both moving continents and atmospheric changes help cause ice ages. Moreover, they say, still other natural forces and

processes may sometimes be involved. One of these processes might be at work in the Atlantic Ocean and other seas. Currents carry warm surface water northward. As this water moves, it steadily cools. When it reaches the northern Atlantic it begins to sink (because cold water is denser than warm water). Then the same water moves southward again along the ocean bottom. In this way, warm and cold waters circulate through the world's seas in what scientists call the **ocean conveyer belt**.

Some researchers suggest that from time to time this giant conveyer belt shuts down. Melting ice releases large amounts of freshwater into the northern Atlantic. Because freshwater is less dense than salt water, it remains on the surface and blocks the flow of warmer water from the south. This cools off the region, perhaps enough to trigger a small and brief ice age.

One theory says this is what happened in the case of some fairly recent cold events commonly called the Little Ice Ages. One occurred between A.D. 1150 and 1460, the other between

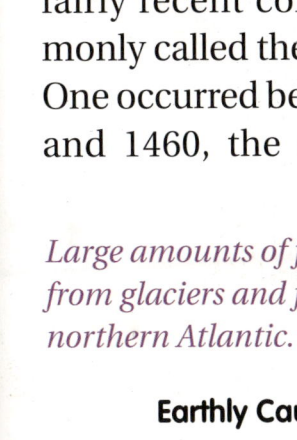

Large amounts of freshwater melt from glaciers and flow into the northern Atlantic.

Earthly Causes for Ice Ages   21

1560 and 1850. Both affected northern Europe and the northern Atlantic region. In the earlier event, ice sheets blanketed Greenland, which had been unusually warm and green for a couple of centuries. This forced the Vikings, who had been colonizing the island, to leave. If they had been able to stay, they likely would have colonized North America next, well before Columbus's voyage. This shows how even a minor ice age can completely change the course of history.

Chapter 3

Cosmic Causes for Ice Ages

Many scientists are convinced that continental drift and decreases of carbon dioxide in the air play roles in starting ice ages. However, they do not rule out the possibility that other factors are also involved. In particular, evidence has been mounting for some time that certain cosmic events may also contribute to the repeating cycle of ice ages. The term **cosmic** derives from the word cosmos, another word for the universe. Thus, cosmic objects and events are those relating to the vast expanse of outer space and Earth's movements within that expanse.

Changes in Earth's Axis

In fact, Earth moves in a number of complex ways within the solar system. One of the ways is Earth's constant **rotation**, or spin, on its **axis**. The axis is an imaginary line or rod running through Earth's

Earth's axis is tilted, causing sunlight to strike some areas of the planet more directly than others.

center from pole to pole. The planet spins around this central rod once every twenty-four hours, defining the day.

What does our planet's axis have to do with climate and the onset of ice ages? First, the axis (and Earth itself) does not stand completely upright. Instead, it is tilted at an angle of about 23 degrees. That means that sunlight strikes the planet more directly in some places than others. And not surprisingly, the areas that receive the more direct rays are warmer, while other areas, especially the poles, are colder.

If Earth's axis were always tilted at the same angle, the present climate might always remain more or less the same. The axis is not always tilted at an angle of 23 degrees, however. Over long periods of time, the tilt grows smaller, then larger, and then smaller again. At one extreme it is just over 21 degrees. At the other it is almost 25 degrees.

At first glance this range of change—about 4 degrees—may not seem very important. But the fact is that a given region of Earth can be several degrees warmer or colder when the tilt is at one of its extremes. And the increase in cold may be just enough to allow glaciers to form. Once they do, these ice sheets can steadily grow larger on their own, perhaps triggering an ice age.

Changes in Earth's Orbit

The idea that alterations in Earth's tilt may be related to ice ages was part of a larger theory proposed by Serbian astronomer Milutin Milankovitch in the 1920s. Milankovitch was convinced that ice ages come about mainly because of cosmic movements and events. In his view, a complex series of planetary movements combine to cause climatic changes, including cold spells and glacial conditions.

In addition to changes in its tilt, Milankovitch said, Earth experiences changes in its orbit. As everyone knows, the planet moves around the sun once each year. This path is not circular, but oval.

Cosmic Causes for Ice Ages

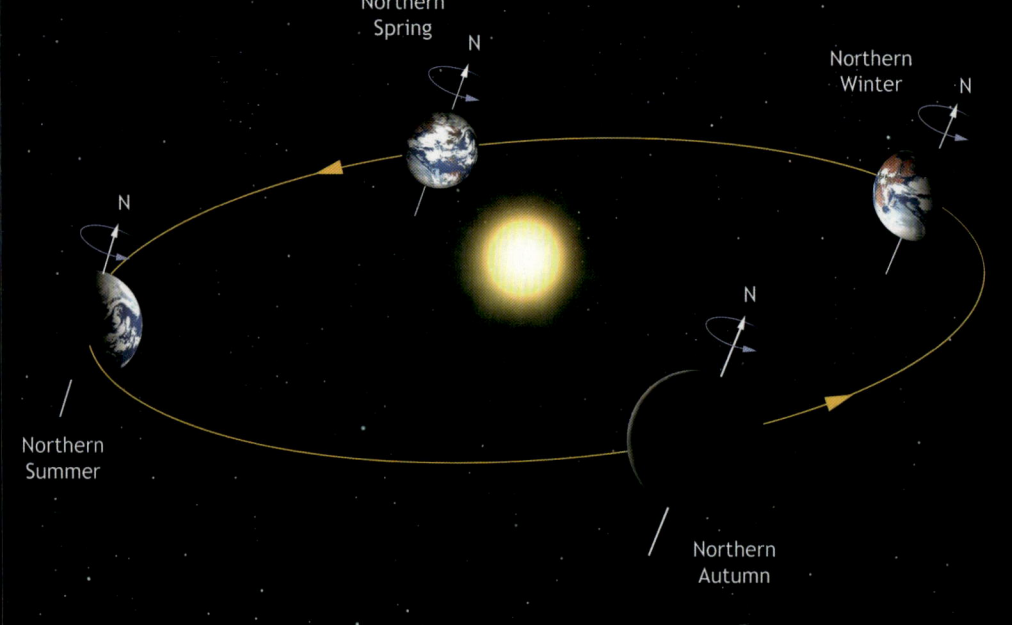

Differences in the amount of incoming sunlight cause the seasons and maybe the advance of glaciers, too.

The technical word for an oval is **ellipse**, so Earth's orbit is said to be elliptical. One obvious consequence is that when the planet lies at the ends of the ellipse it is slightly farther away from the sun than when it lies at the sides of the ellipse. (If the orbit was circular, the distance to the sun would always be the same.)

Complicating matters is the fact that Earth's elliptical orbit changes slightly in shape over long periods of time. Sometimes the oval is slightly longer and thinner than it is at other times. The degree of change in the ellipse's shape is called **eccentricity**. Very long and thin ellipses are highly eccentric, while ellipses that are closer in shape to circles are only slightly eccentric.

The point is that Earth's orbit can vary in eccentricity by up to 5 percent. And according to Milankovitch, when the elliptical orbit is most eccentric the planet can receive less sunlight. This, he said, combined with differences in tilt and other factors, might cause an ice age about every 100,000 years. Scientists have come to call this period and process a Milankovitch cycle in his honor.

Other Cosmic Theories

A majority of researchers agree that Milankovitch cycles do roughly correspond with the occurrence of several of the known ice ages. The theory does

The formation of glaciers like this one in Alaska may be influenced by changes in Earth's orbit.

Comets, which contain large amounts of ice, strike Earth from time to time.

not explain all of the ice ages, however. So scientists continue to search for alternate causes, including cosmic ones, for these unusually cold eras.

Some of the recent cosmic theories have been quite colorful. One proposal suggested that on occasion, Earth moves through an abnormally cold region of space. And when it does, the climate in some parts of the world grows cooler, promoting the growth of glaciers. Many scientists are wary about this theory because it is hard to tell which areas of outer space are colder than others. Extensive exploration of the solar system, taking decades or perhaps centuries, will be needed to collect the data to prove or disprove the theory.

Another explanation for ice ages was suggested in the 1960s by D.W. Patten. He proposed that large amounts of space ice shower Earth from time to time. (Patten also tried to link such ice falls to the biblical flood.) However, a majority of scientists have rejected this theory, saying that most of the incoming ice would melt as it sped downward through the atmosphere.

Much more promising is a theory proposed in the late 1990s by scientists Richard A. Muller and

As Earth moves through space, it sometimes encounters space dust, which may cool the planet's surface.

Gordon J. MacDonald. They say that from time to time, large clouds of space dust cross Earth's orbit. When the planet passes through a dusty region, many of the particles enter the upper atmosphere, where they remain for a while. During this interval, the dust blocks sunlight just enough to cool some regions of the planet's surface and trigger an ice age.

It is possible that the process Muller and MacDonald describe did contribute to at least some of the ice ages. If so, it probably worked in concert with other factors, including earthly ones such as continental drift. In fact, it is likely that no single event or process caused all the ice ages. At present, a majority of scientists agree that one combination of factors caused some ice ages and a different group of factors caused others. This greatly increases the difficulty of predicting any future ice ages.

Chapter 4

Global Warming and Future Ice Ages

By about 10,000 years ago, the ice sheets from the last large-scale glacial period had receded from many parts of the Northern Hemisphere. This gives the false impression that the great ice age cycle that began nearly 3 million years ago is over. Indeed, many people are under the mistaken impression that ice ages in general are a relic of the past.

In reality, scientists agree that the ice ages are not over. Large ice sheets and snowpacks still blanket the polar regions, Greenland, and other areas. And only when the majority of these frozen wastes are gone will the present ice age cycle be over. The fact is that Earth and humanity are merely enjoying the relative warmth of the latest interglacial period. There is no doubt that eventually a new glacial event will engulf large parts of Europe and North America.

The two crucial questions involved are: When will this happen? And how will it happen? Regarding the second question, there is reason to believe that, for

Many glaciers are now retreating. But they will eventually return, starting a new ice age.

32 The Ice Ages

the first time in history, the causes will not be entirely natural. Evidence suggests that humans may already be speeding up the onset of a new ice age. Moreover, as strange as it may sound, people might end up making the world colder by first making it warmer.

Global Warming

Indeed, some scientists worry that the trigger for the next ice age may be the phenomenon known as **global warming**. The term refers to rises in the temperature of Earth's atmosphere brought about by human activity. More specifically, that activity is related to heavy industry and energy production,

Global Warming

❶ The Sun releases energy in the form of light and heat. About 70 percent of the Sun's heat is absorbed by land, air, and the oceans.

❷ About 30 percent of the Sun's heat reflects off the Earth's surface and atmosphere back into space.

❸ Some of the reflected heat is trapped by greenhouse gasses in the Earth's atmosphere. Global warming occurs when high levels of greenhouse gasses trap more heat.

including the burning of fuels such as oil, gasoline, and coal. Such burning and many industrial processes create various unwanted by-products. Some, like soot, are solid. But others are gases, including carbon dioxide, nitrous oxide, methane, and ozone. It has been established that carbon dioxide has the ability to absorb and trap heat from sunlight. Nitrous oxide, methane, ozone, and a number of similar gases do the same. Scientists call these **greenhouse gases** because their heat-trapping effect is similar to that produced by the glass ceiling and walls of a greenhouse.

Earth's Temperature Increase

The Earth's temperature has increased significantly since the industrial revolution in the 1800s. It has increased by half a degree in the past twenty-five years alone.

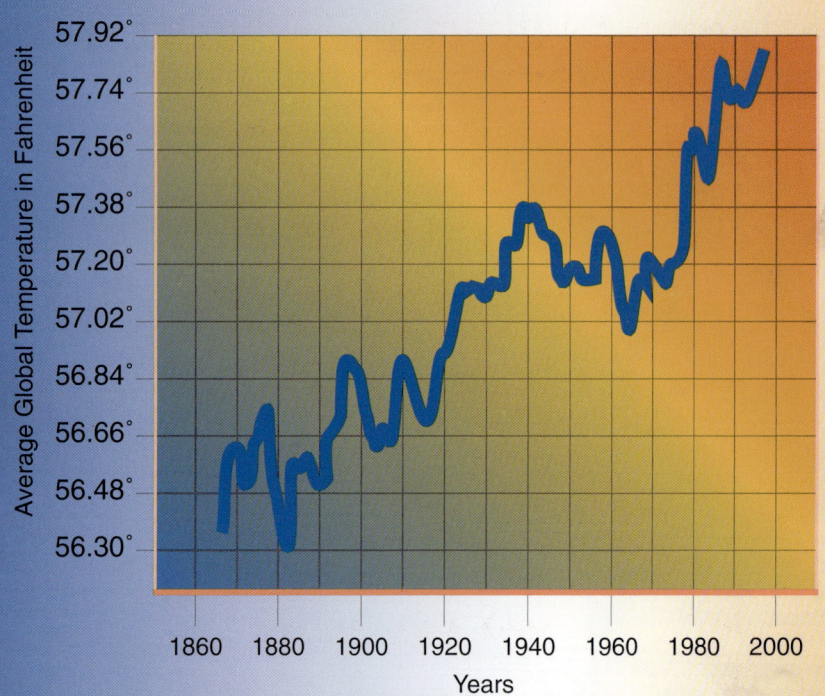

The problem is that these greenhouse gases are steadily increasing the temperature of the atmosphere's lower levels. Consider carbon dioxide levels alone. Experts estimate that before the industrial revolution (which began in earnest in the early 1800s), the level of carbon dioxide in the air was about 280 parts per million (ppm). (In other words, for every million molecules of air there were 280 molecules of carbon dioxide.) Today, carbon dioxide levels average 370 ppm. And by the end of the present century, these levels will likely reach at least 500 ppm. Levels of ozone and other greenhouse gases are also increasing.

The effects of this warming are already apparent. The overall atmospheric temperature has increased by half a degree in the past 25 years alone. Mountain glaciers in many areas of the world are melting at alarming rates. And warmer waters in parts of some of the oceans have begun to generate more frequent and stronger storm systems in some regions.

The Northern Atlantic at Risk?

It is only logical to ask how such warming could cause the opposite—climatic cooling—and maybe even trigger an ice age. At first glance, this idea certainly seems very *illogical*. But first one must realize that the term global warming is often largely misunderstood. Many people assume that it means that all parts of the world will grow warmer and eventually experience constant summer.

The reality is that increases in temperature produce different climate reactions in different areas. This is partly because of the planet's tilted axis. The warmth from sunlight is never evenly distributed to all the continents and seas. Similarly, when global warming causes some areas to grow warmer, other regions will grow colder to compensate.

Much more important, however, is the effect that global warming may have on the ocean conveyer belt. In particular, a number of scientists warn, the portion of the belt in the northern Atlantic may be at risk. The fact is that the temperature of these waters affects air temperatures and the weather in nearby continents. Constant circulation of warm and cold waters within the North Atlantic helps to maintain flows of warm air over North America, Europe, and several other regions.

If global warming continues, portions of the ice sheets in Greenland and the Arctic will melt. Large amounts of cold freshwater will flow into the North Atlantic. This could block the normal flow of warm water in that ocean and shut down the ocean conveyer belt. The result would be a rapid cooling effect that could promote the growth of large ice sheets in the Northern Hemisphere.

In fact, some convincing recent evidence suggests that at least a minor ice age could occur by the end of this century. And a number of researchers think it might happen a good deal faster. Among them are members of the prestigious

National Oceanic and Atmospheric Administration (NOAA). In a March 18, 2004, article in the *Christian Science Monitor*, writer Peter N. Spotts quotes Mark Eakin, a leading NOAA scientist, as saying, "There's the very real potential of the climate system changing dramatically and rapidly."

Global warming might cause some of Greenland's ice sheets, seen here, to melt.

Planning for the Future

It should be emphasized that neither Eakin, his colleagues, nor anyone else is 100 percent certain this scary event will definitely happen. But the possibility has become real enough to make politicians and national leaders take notice. At the urgings of NOAA scientists, members of the U.S. Senate's Commerce, Science, and Transportation Committee considered the problem early in 2004. In March of that year, they drafted a bill giving the NOAA $60 million. The money was intended to fund further

A group of U.S. senators visits an island in the Arctic region to study the effects of global warming.

research into the possibility of sudden climatic change, especially the onset of colder conditions. Meanwhile, the U.S. Department of Defense has also begun to investigate the problem.

These government groups, along with experts at the NOAA, hope to achieve two challenging goals. First, they want to find ways to predict major climatic changes more accurately. This will allow people to prepare better for any impending disaster. Second, those studying the problem want to formulate strategies for slowing global warming. This, they hope, will greatly reduce the possibility that humans will trigger an ice age.

However, even if this goal is met and humans do not cause a new ice age, they will succeed only in delaying the inevitable. They may be able to avoid the ice sheets for a century or two, or perhaps for a few thousand years. But sooner or later, nature will have its way. Scientists are almost certain that more ice ages will occur in the future through traditional natural causes. As in the past, large areas of North America and Europe will be covered by ice sheets. Many cities, towns, and farms that are comfortable and prosperous today will have to be abandoned.

Yet the human race need not despair. People with little more than stone tools and animal furs at their disposal managed to survive several glacial periods. Present and future technology will ensure that humanity will survive even the coldest ice age. It is

A scientist at the South Pole launches a weather balloon to study temperatures in the upper atmosphere.

even possible that an advanced technology no one has yet dreamed of will someday allow people to halt and reverse ice ages. Only time, of which nature has an unlimited supply, will tell.

Glossary

axis: An imaginary line or rod that runs through an object's center and around which the object spins.

carbon dioxide (or CO_2): A common gas absorbed for nourishment by plants but expelled as a waste product by animals and people. Carbon dioxide is also produced by some industrial processes and by burning coal and oil.

cosmic: Having to do with things beyond Earth or in outer space.

eccentricity: The degree to which an ellipse gets more elongated (more eccentric) or closer in shape to a circle (less eccentric).

ellipse: An oval shape.

glaciers: Rivers of ice, often hundreds or thousands of feet thick, that form in cold regions and can expand and move slowly across the land.

global warming: A current scientific theory that claims that a combination of natural and human factors is causing an increase in temperature in the lower levels of Earth's atmosphere.

greenhouse gases: Gases that can absorb and trap heat, including water vapor, carbon dioxide, ozone, and methane.

interglacial: A period of relative warmth that separates one ice age or glacial period from another.

NOAA (National Oceanic and Atmospheric Administration): Founded in 1970, this U.S. government agency brought together some of the country's oldest and best scientific research groups, including the Weather Bureau, established in 1870.

nomadic: Moving from place to place, as in the case of people who migrate to follow herds of animals that they depend on for food.

ocean conveyer belt: A system or process in which warm and cold water circulates through the world's oceans. The process is most pronounced in the northern Atlantic Ocean.

plate tectonics: The science that examines the slow movements of the continents across Earth's surface.

precipitation: Falling moisture, including rain, snow, drizzle, hail, and ice.

rotation: Spinning, usually on an axis.

For Further Exploration

Books

Karen J. Donnelly, *Ice Ages of the Past and Future.* New York: Power Kids Press, 2003. An easy-to-read overview of the ice ages of the past and speculation about future ice ages.

Ian Lange, *Ice-Age Mammals of North America.* Missoula, MT: Mountain Press, 2002. This introduction to the wooly mammoth and other creatures that roamed during the recent ice ages is colorfully illustrated.

Nick Merriman, *Early Humans.* London: Dorling Kindersley, 2000. An excellent introduction to Stone Age humans, including reconstructions of their hard life in ice age conditions.

Fred Pearce, *Global Warming.* London: Dorling Kindersley, 2002. Provides a summary of the present evidence for global warming and discusses the possibility that it might help trigger a new ice age.

Web Sites

The Big Chill (www.pbs.org/wgbh/nova/ice/chill.html). An introduction to the ice ages, presented by the prestigious NOVA television production team.

Glaciers: Rivers of Ice (http://members.aol.com/scipioiv/glmain.html). Tells about how glaciers formed and moved during the ice ages.

Ice-Age Mammals (www.zoomdinosaurs.com/subjects/mammals/iceagemammals.shtml). A terrific site that gives lots of information on, plus color drawings of, all of the major creatures that lived during the recent ice ages. Highly recommended.

Index

Antarctica, 18
Arctic Circle, 18
atmosphere, of Earth,
  18–20, 33–36
axis, of Earth, 23–25, 36

Canada, 9–10
carbon dioxide (CO_2),
  18–20, 34, 35
causes
  changes in atmosphere,
    18–20
  changes in axis, 23–25
  changes in orbit, 25–27
  movement of continents,
    15–18
  of next ice age, 33–36
  ocean conveyor belt,
    21–22
  from outer space, 28–30
  sunlight, 13, 15, 18
Charpentier, Jean de, 5
Christian Science Monitor
  (newspaper), 37
climate
  axis of Earth and, 24
  sunlight and, 13, 15, 18
continental drift, 15–18
Cryogenian ice age, 6–8

Eakin, Mark, 37
eccentricity, orbital, 26
Elbe glacial period, 8

Elster glacial period, 8–9
equator, 17–18

folk tales, 4–5

glaciers
  growth of, 13, 15
  movement of, 4, 6–8
global warming, 33–36, 39
greenhouse gases, 34–35
Greenland, 22, 36

humans, during
    Pleistocene ice age, 12

Illinoian glacial period, 9
interglacial periods, 8, 31

Kansan glacial period, 8–9

Laurentide, 9
lions, 12
Little Ice Ages, 21–22

MacDonald, Gordon J., 30
major ice ages, 6–12, 18
mammals, 10–12
mammoths, 12
megaceros, 11–12
Milankovitch, Milutin, 15,
  17
movement
  of continents, 15–18

of glaciers, 4, 6–7
Muller, Richard A., 29–30

National Oceanic and Atmospheric Administration (NOAA), 36–39
Nebraskan glacial period, 8
next ice age
 causes of, 33–36
 predicting occurrence of, 36–39
North Atlantic Ocean, 35–36

ocean conveyor belt, 21–22, 36
orbit, of Earth, 25–27
outer space, 28–30

Patten, D.W., 29
plants, 19–20
plate tectonics, 16–18
Pleistocene ice age, 8–12, 18

polar regions, 18, 24, 36

runaway cooling, 15

Saale glacial period, 9
space dust, 30
space ice showers, 29
Spotts, Peter N., 37
sunlight
 axis of Earth and, 24
 carbon dioxide and, 18–19, 34
 climate and, 13, 15, 18
 dust clouds and, 30
 orbit of Earth and, 27

United States, 9–10

Vikings, 22

Wegener, Alfred, 15–16
Weishsel glacial period, 9–10
Wisconsinan glacial period, 9–10

Picture Credits

Cover: PhotoDisc
The Art Archive/National Anthropological Museum Mexico/Dagli Orti, 11
© Tore Bergsaker/Corbis, 38
Corel, 5, 16, 32
Bernhard Edmaier/Photo Researchers, Inc., 37
Mark Garlick/Photo Researchers, Inc., 26, 28
Joseph Giddings, 19
Gary Hincks/Photo Researchers, Inc., 7, 9
Joyce Photographics/Photo Researchers, Inc., 40
© Danny Lehman/CORBIS, 27
David Mack/Photo Researchers, Inc., 29
© Gianni Dagli Orti/CORBIS, 10
© Paul A. Souders/CORBIS, 20
Simon Terrey/Photo Researchers, Inc., 6
Detlev Van Ravenswaay/Photo Researchers, Inc., 24
Frank Zullo/Photo Researchers, Inc., 14

About the Author

In addition to his acclaimed volumes on ancient civilizations, historian Don Nardo has published several studies of modern scientific discoveries and phenomena. Among these are *The Extinction of the Dinosaurs, Atoms, The Solar System, Black Holes*, and a biography of Charles Darwin, who advanced the modern theory of evolution. Mr. Nardo lives with his wife, Christine, in Massachusetts.